Rafael Galli
Igor Barros
Gustavo Kuhn

WHEELCHAIR CONTROLLED BY VOICE AND BLUETOOTH

Rafael Galli
Igor Barros
Gustavo Kuhn

WHEELCHAIR CONTROLLED BY VOICE AND BLUETOOTH

Development of a wheelchair controlled by voice command and Bluetooth

ScienciaScripts

Imprint

Any brand names and product names mentioned in this book are subject to trademark, brand or patent protection and are trademarks or registered trademarks of their respective holders. The use of brand names, product names, common names, trade names, product descriptions etc. even without a particular marking in this work is in no way to be construed to mean that such names may be regarded as unrestricted in respect of trademark and brand protection legislation and could thus be used by anyone.

Cover image: www.ingimage.com

This book is a translation from the original published under ISBN 978-620-2-80709-8.

Publisher:
Sciencia Scripts
is a trademark of
International Book Market Service Ltd., member of OmniScriptum Publishing Group
17 Meldrum Street, Beau Bassin 71504, Mauritius
Printed at: see last page
ISBN: 978-620-3-48560-8

VOICE-CONTROLLED WHEELCHAIR
E
BLUETOOTH

By Gustavo Peglow Kuhn[1], André Wille Lemke[2], Luiz Otávio Victória[1], Thales Gonçalves Ferreira[1], Igor da Rocha Barros[3], Léo dos Santos Ribeiro[3] & Rafael Galli[3]*.

[1] *Electrical Engineering students, IFSUL Câmpus Pelotas, 55740-000, Pelotas, Brazil.*

[2] *Electrical Engineering students, UFPel, Pelotas, Brazil.*

[3] *Professor of the Technical Course in Electronics, IFSUL Câmpus Pelotas, 55740-000, Pelotas, Brazil.*

*E-mail: rgalli@pelotas.ifsul.edu.br

Pelotas - Rio Grande do Sul - Brazil
2019

SUMMARY

This project aims to help designers or researchers in the field, or people with physical disabilities, to obtain better equipment to improve the comfort of wheelchair users. The project was carried out after researching the number of people who need to use wheelchairs or depend on others to get around. With this research on paraplegic and quadriplegic people in hand, ways to improve the quality of life of these people were studied, with the idea of using motorized wheelchairs that are controlled by voice or by Bluetooth. The results of this research will be presented in this work, including details of the electrical circuits and the source code of the voice control stage, of Bluetooth, and of the Android application.

Keywords: Motorized wheelchair. Voice control. Assistive technology.

Thank you

Before starting the presentation of the project, we would like to thank everyone who guided and supported us so that this work could become a reality. First of all we would like to thank our family members for always supporting us in all aspects, and understanding our absences during the project development process.

We thank the company NOBRESOL Soluções em Sistemas de Energia in the person of Jorge Luiz Teixeira da Costa, for the financial and technical support, and the company IMOBRAS for the engines used in the construction of the full-scale prototype.

To all the students and teachers that make up Laboratory 14 Research and Development for the support in difficult times. To the servers and teachers of the IF-Sul Campus Pelotas in special to the Technical course in Electronics where Laboratory 14 is located and to teachers Gladimir Pinto da Silve and Frederico Grequi, from the Technical course in Mechanics, for the support and guidance in the calculations of mechanical structures. To all those mentioned above and others who are somehow involved in the construction process of this work, we can only say "*Thank you very much*".

1 - Introduction

What is "Voice Recognition" and how does it work?

Speech recognition is an advanced technique that analyzes acoustic waves captured by a microphone by looking for patterns in those waves. When the user says something, the system compares the sequence of waves captured with the patterns saved in its database. When the sequence spoken by the user matches one in its database, it performs a pre-programmed action, such as turning on an LED or walking forward, in our case.

2 - Development

The entire development of the project was carried out in the Research and Development Laboratory 14, of the Electronics technical course, of the Instituto Federal Sul-Rio-Grandense, Pelotas campus, Brazil.

Founded in 2006 with the goal of providing a place for students to put their ideas into practice. Ideas that have already won fairs and have been known nationally and internationally.

The challenge proposed here is teaching-learning, guided by the principles of educational theorists, which are based on doing in order to learn (Maria Montessori,2014).

2.1 - General Objective

This project had as its objective the creation of a Voice-Controlled Wheelchair with the intention of providing a better quality of life for people with special needs. It also contributes to the country's technological advance in terms of wheelchair manufacturing, making it possible for the device to be implanted in existing wheelchairs.

In addition to developing an Android application, to control the chair from a distance, allowing the user or relative to control the chair from a distance.

2.2 - Specific Objectives

Acquire knowledge in electronics and software development for embedded systems;
Learn more about how to write reports, logbooks, and so on.
Achieve perfect voice command operation, so that whatever the user speaks, the chair obeys his or her commands.
Minimize the cost of the project.

3 - Operation

The block diagram in figure 1, shows the logical sequence of the electronics of the chair and will be presented block by block below.

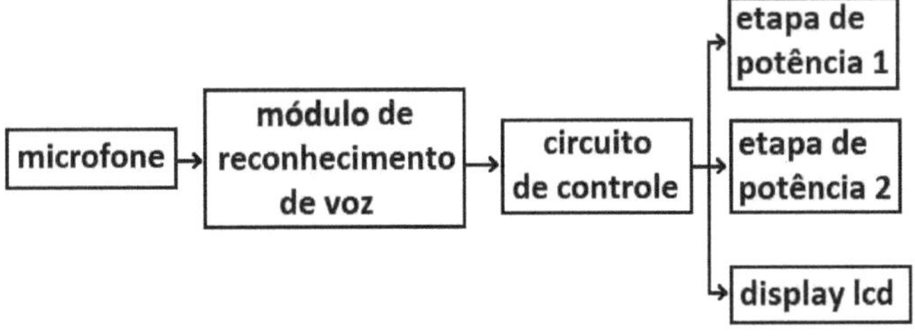

Figure 1- Block Diagram

4 - Microphone

The microphone picks up the pressure variations (sound waves) on its membrane and transforms them into low-amplitude electrical signals, about 100mV. The ideal frequency range of a microphone would be from 20 to 20000 Hz, the same as a human being, but common microphones are from 200 to 2500 Hz.

Figure 2, shows an example of a Microphone, the same one that was used in our experiments.

Figure 2- Electret Microphone

5 - Voice Recognition Module

The voice recognition module receives the electrical signals from the microphone and compares them with the commands that the user has saved in its database. If the signal received is equivalent to one of the saved patterns, the module sends the control circuit a number for that command. The communication between the module, seen in figure 3, and the control circuit is serial, using the "RS232" standard.

Figure 3- Voice Recognition Module

6 - Control Circuit

For the control part a PIC 16F877A microcontroller was used. This was chosen for being the lowest cost among those available at the institution. It was able to process all the data, without generating a bug due to process overload. A 20MHz crystal clock was used. The schematic of the board can be seen in figure 4 along with the PCB and the prototype, figures 5 and 6, and the list of components is in appendix c.

The control circuit generates two PWM (pulse width modulation) signals according to the data it receives from the voice recognition module, via serial communication. These PWM signals are sent to the power boards, which will drive the DC motors.

To avoid "jerking", a ramping algorithm has been implemented via software. This algorithm gradually increases or decreases the PWM pulse width (topic 8 explains what PWM is), thus smoothly accelerating or decelerating the motors.

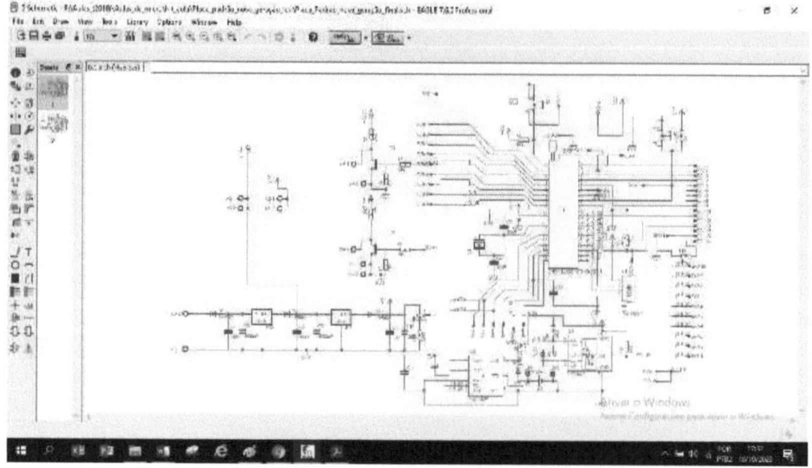

Figure 4- Control Board Schematic

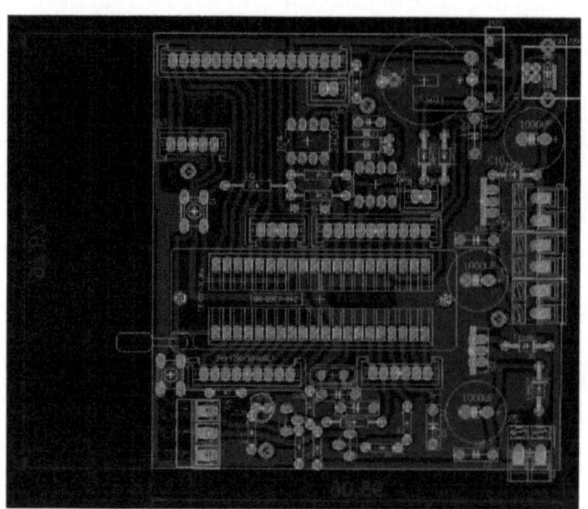

Figure 5- Control Board PCB

Figure 6- Control board prototype

8 - PWM

PWM stands for Pulse Width Modulation. It is a power control technique that is based on switching the load on and off at each preset time interval.

The basic operation of a *PWM* will be demonstrated using the graph in figure 7.

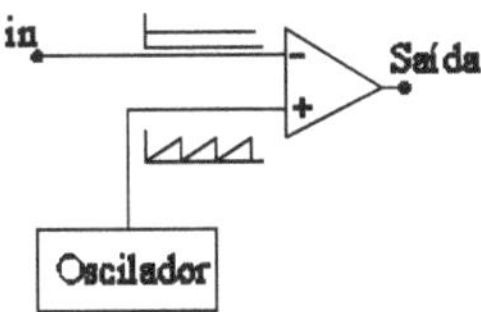

Figure 7- Demonstration Circuit.

A designer-defined sawtooth signal of constant period T is injected into one of the comparator inputs.

At the other terminal we enter a "modulating" signal that can be a direct voltage or any other signal.

In the following example, we will use a direct voltage for ease of understanding, divided into three cases that can be seen in the graphs in figure 9.

In the first case, case 1, a DC voltage, approximately half the ramp voltage,   is injected into the input In, the *PWM* output remains at high level until the ramp crosses with the voltage injected into the input In, when the two voltages cross, the *PWM* output drops to zero remaining in this state until the reference ramp drops to zero forcing the output to return to the initial state (high level).

In the second case, case 2, a voltage is injected into *In with* the value approximately equal to the maximum amplitude of the sawtooth wave, it is noted that the width of the output pulse is increased until the two signals, sawtooth and DC voltage intersect, returning the output to the normal state after the sawtooth falls to logic level zero.

Finally in case 3, we inject a signal close to zero, which generates a small high level output.

What we can see, observing these three cases is that depending on the position where the input   signal *in is*, we will have a proportional variation of the output always within the same period T, figure 8.

This variation is called the Cyclic Ratio and is defined by:

$$D = \frac{T1}{T}$$

Figure 8- Cyclic Ratio Equation

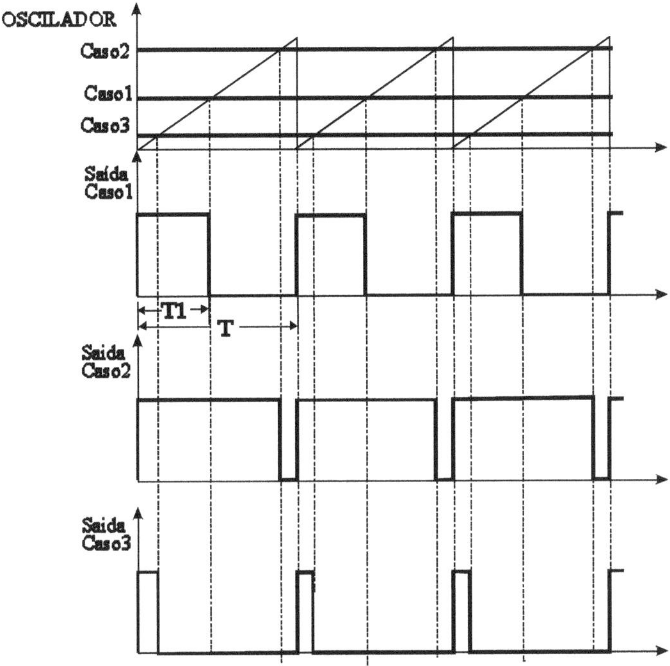

Figure 9- Chart demonstrating how PWM works

Its efficiency is much higher than linear control, because the transistors work only in the cut-off or saturation region. They behave as open or closed switches.

The high efficiency is explained mathematically. First, power equals the product of voltage and current ($P = V \times I$). To make a comparison we will use a 20V, 10A, 200W motor. When the transistor is in the cut-off region, the current tends to zero. The product of any number with zero is equal to zero. Now, when the transistor is working in the saturation region, the voltage across its terminals is about 0.3V (it can vary a bit according to the type of transistor used). So the power will be equal to a little less than a third of the current. In this case only 3W. In other words, the total power (of the motor plus the loss) is 203W, and the efficiency is 98.52%.

9 - H Bridge

The H-bridge is a transistor (or relay) arrangement for driving DC motors (direct current motors), in our case. This arrangement allows reversing the polarity of the motor supply, which reverses the direction of rotation of the motor.

Usually the H-bridge is driven by a PWM signal. This allows, besides the control of the direction of rotation, the control of the motor speed.

The complete circuit of a Full Bridge Inverter is shown in Figure 10. In the circuit we put a transformer to represent our load.

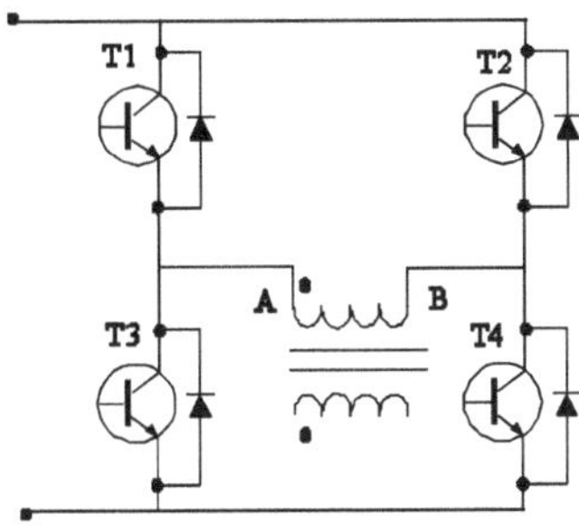

Figure 10- H Bridge

At first T1 and T4 are driven by biasing the output transformer with plus (+) at point "A" and minus (-) at point "B", as shown in figure 11.

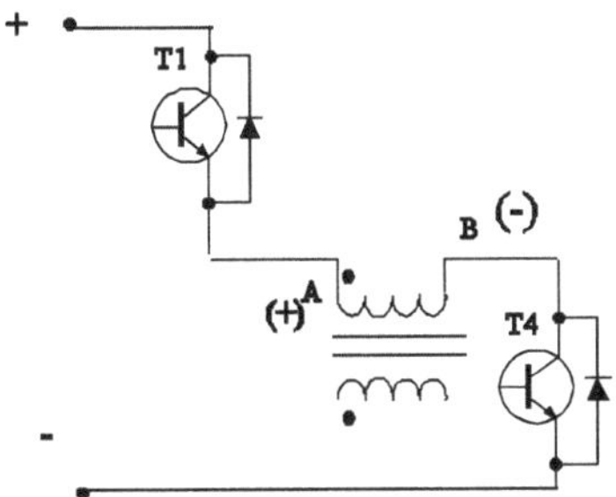

Figure 11- 1st stage of the H-bridge operation

In a second moment, we have the deletion of T1 and T4. In a 3rd moment T2 and T3 are activated, now polarizing point "A" with minus (-) and point "B" with plus (+), as we can see in Figure 12. And to finish in a 4th moment T2 and T3 are turned off and so on.

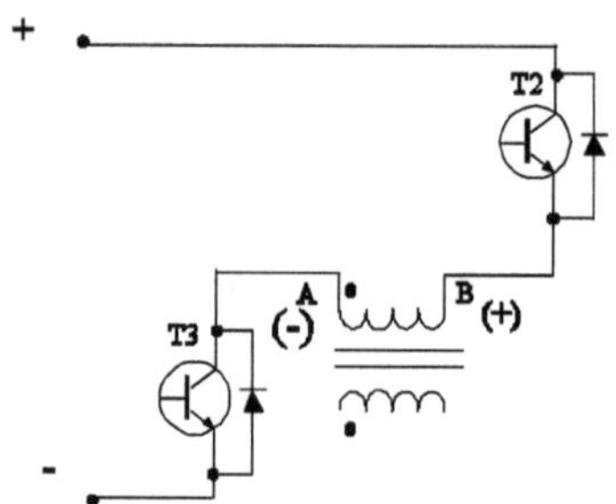

Figure 12- 3rd step of the H-bridge operation

In this way the waveform represented in Figure 13 is obtained.

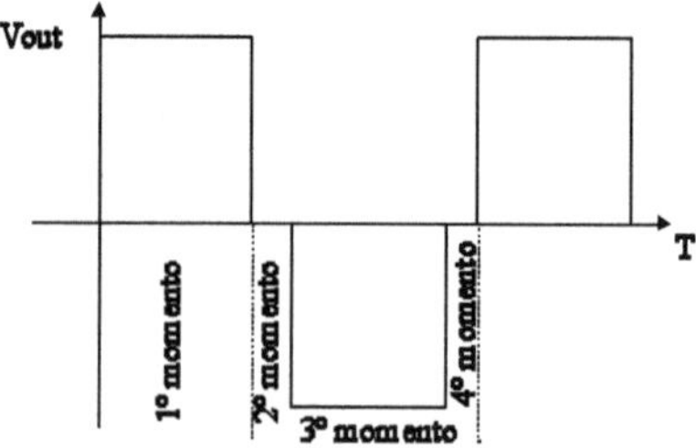

Figure 13- Output waveform and its drive steps

10 - LCD Display

The *display* will show information for better communication with the user, such as, for example, the commands received by the voice recognition module and the charge in the batteries, in figure 14 we show an example of LCD.

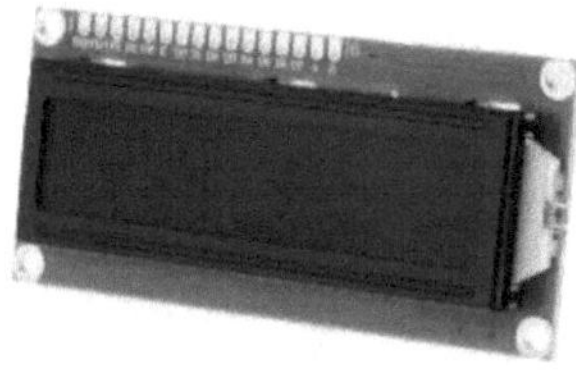

Figure 14- LCD

11 - Power Stage

For the power stage two H-bridge output boards were used (topic 9 explains what an H-bridge is). Each board controls a motor, allowing us to move the chair to the left or to the right and the fact of being H-bridge, allows the motors to have clockwise or counterclockwise rotation, according to the signal coming from the control board, enabling the control of the chair in all directions.

The schematic circuit that is shown in Figure 15 shows the complete circuit of the power stage, in this circuit can be observed two integrated circuits of 8 pins, these are two power drivers of the Internationl Rectifier IR 2111, whose function is to drive the MosFet's power, in a simple way, reducing the number of components of the board lowering the cost of the project.

These drivers have a number of protections, such as the high MosFet's never being triggered with the low MosFet's, that make the circuit safer.

During the research for the development of the power board, we found that the Internationl Rectifier, had another eight-pin integrated, IR 2184, with very similar characteristics, but with better internal peripherals, including Shutdown, which allows the shutdown of the power stage leaving the motors "free".

In this way we can, through a set of jumpers, select which integrated will be used as the power driver.

Figures 15 and 16 show the configurations for each of the integrated chips.

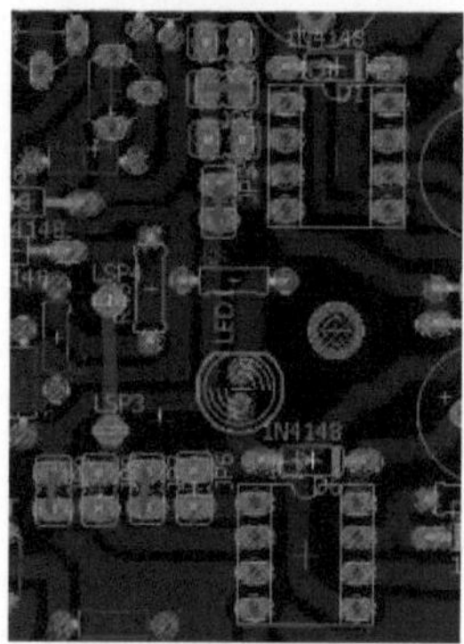

Figure 15- Setting the Jumpers for IR 2184

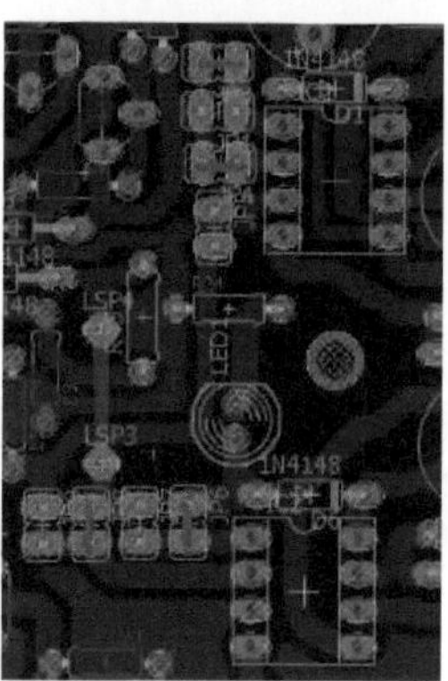

Figure 16- Setting the Jumpers for IR2111

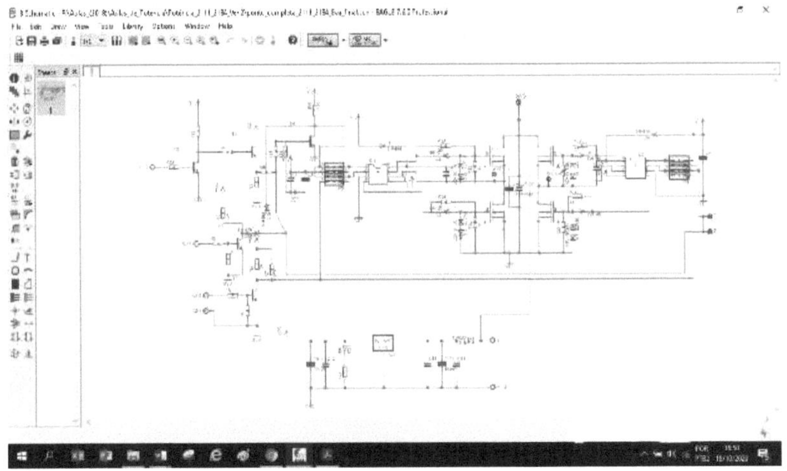

Figure 17- Electrical diagram of the H-bridge

The circuit was designed in Eagle 7.6.0 software and the PCB board designed in the same software can be seen in figure 18 the component list is in appendix D.

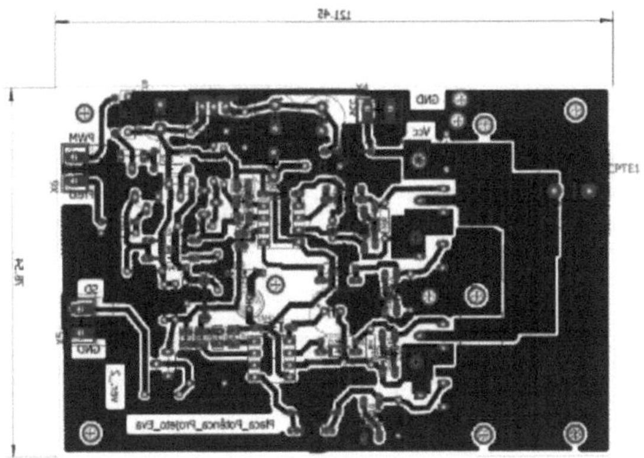

Figure 18- Power stage PCB

This board has a PWM input, an electrical brake input, that when it receives a high level, it turns on the low MosFet's of the H-bridge, generating a short on the motor, braking it. And for the version with IR 2184, it has a Shutdown control pin that turns off the H-bridge MosFets.

Figure 19- Power Board prototype

In figures 20 and 21 we can observe the signals taken from the prototype Power board, showing the stability of the signals,

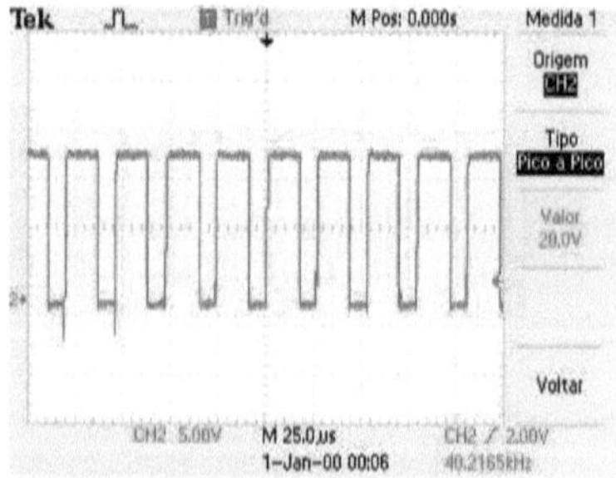

Figure 20- PWM at the Power Drivers input

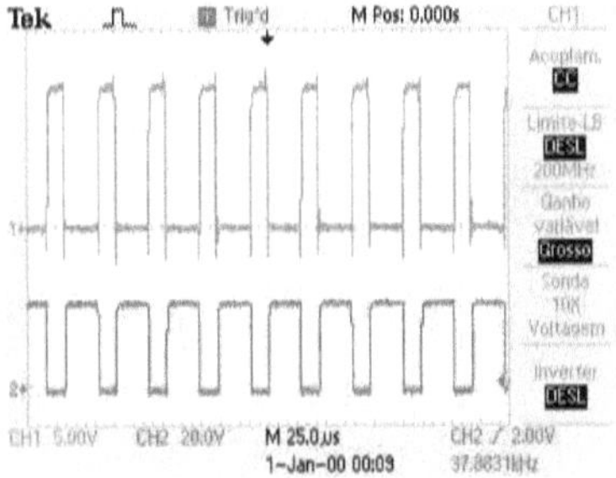

Figure 21- Output Signal from Power Drivers

The signals were obtained with the aid of a Tectronix TDS 2022B oscilloscope and the Open Choice PC Communication software, also from Tectronix.

On this board, protection circuits were also provided for the Fet gates to prevent them from being damaged by switching noise.

During the research, of the power stage, it was found that the gates do not support voltages higher than 20V, hence the need for such a protection circuit, since the switching noise far exceeds this voltage.

The circuit used, for the protection, is recommended by the manufacturer and can be seen in Figure 22. It is important to emphasize that this protection must be done in the four MosFet's of the H-bridge.

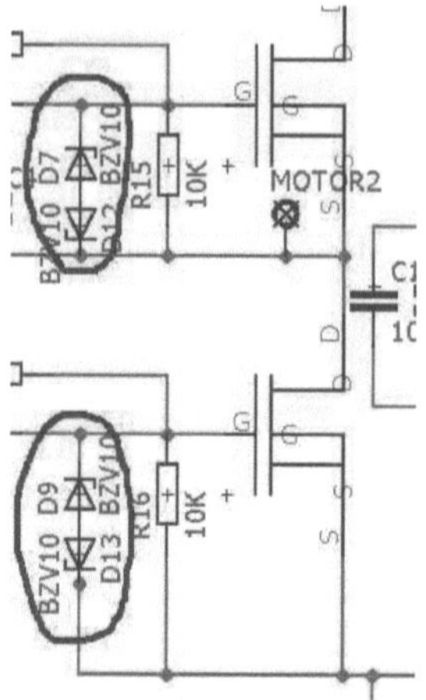

Figure 22- MosFet's Gate Protection Circuit

11 - Bluetooth Communication

During the development of the project, the need was seen for an alternative control, the remote control of the chair to perform evaluative tests on the control circuits while the voice command was not working.

For this task a Bluetooth module was used, and an application for control was developed,

We also had the idea of implementing this type of communication, in the final prototype, because we found, throughout our research, that some chair users had difficulty accessing their chairs when, for some reason, they were out of reach.

The application, developed for testing, mentioned above, has become a good solution to this problem, allowing the user   to access his chair using a cell phone with the Android platform.

11.1 - What is Bluetooth

Bluetooth is a form of communication developed by the Ericsson company in 1994, this system allows wireless data exchange.

It was named after a former king of Denmark and Norway, Harold Blatand, which in English is spelled, Harold Bluetooth.

The system transmits on a 2.4GHz radio frequency, most devices have a range of 1 to 100m.

The module used for the project was Arduino's HC-06, which can be seen in figure 23.

Figure 23- Arduino Bluetooth Module

This module converts the data received via radio wave and transmits it to the control board via serial, which makes it much easier to work with.

12 - Mechanical Structure

For the field tests, two prototypes were created. The first was a miniature to analyze the embedded electronic circuits, mainly the control board and the voice acquisition board, the power boards were made using ULM 2003, because stepper motors were used due to the prototype's scale of 1:4, the prototype can be seen in figure 24, this was thought in such a way as to make it possible, in the future, to test other forms of control of the chair, such as an automaton system for people with severe injuries.

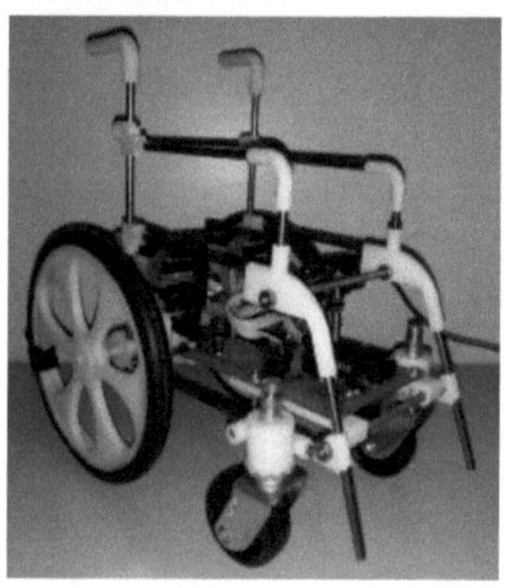

Figure 24- Prototype for bench test in 1:4 scale.

The structure was made with metallic wire and the junctions in ABS plastic and printed on a 3D printer model MOJO. Later the prototype of the figure 25, was replaced by the figure 25, because it allowed a better accommodation of the electronic circuits and sensors, as previously mentioned, for future modifications.

Figure 25- New 1:4 scale prototype

In this prototype a new power board was developed, because due to the weight of the battery, new, more powerful stepper motors were needed.

The power board for the second prototype (scale 1:4) has eight NPN transistors (TIP122), capable of switching the coils of the motors, making their movement possible. The 74HC595 integrated circuit was used in order to reduce the number of wires and connections between the main board and the power module.

The field tests were performed on a chair structure developed by the laboratory students and built with the support of the institution's locksmith shop. It is basically a common motorized wheelchair, with two 24V motors, 4 poles Universal, 280W, 2800 rpm model 101340124 from the company IMOBRAS, which lent us the motors for the construction of the full-scale prototype. It was decided to work with 24V so that the currents involved were lower than those of 12V motors.

The motor also presents a very interesting physical characteristic, besides being completely sealed, making it difficult for water or dust to enter, its housing has four fixing holes, which can be seen in figure 26, which makes assembly much easier. The image of the motor can be seen in Figure 27.

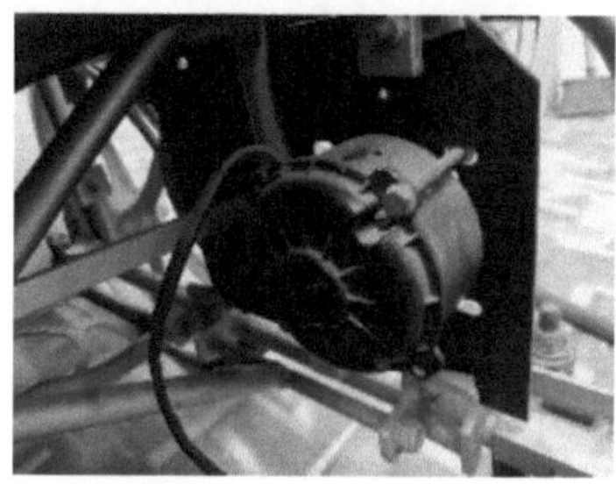

Figure 26- Attaching the motor to the chair

Figure 27- 24V Motor

As the torque on the shaft of the motors are usually low, it was necessary to make two torque ratios, which also allowed us to calculate the maximum speed of the chair.

12.1 - Calculating the chair's final speed

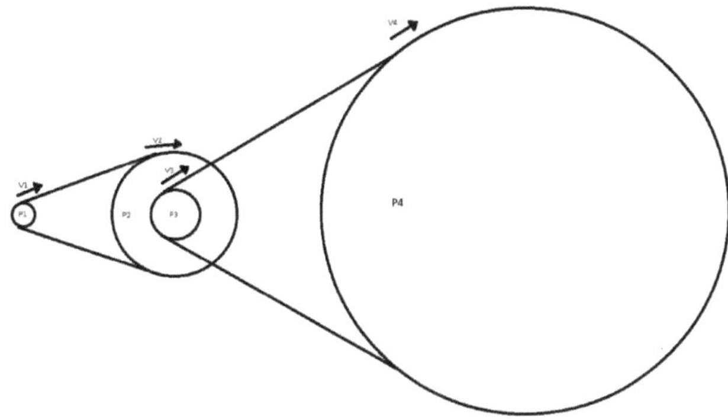

Figure 28- pulley ratio.

To perform the calculations of the chair's final speed, some data were needed from the motor, data that is represented in figure 29, and from the pulleys in figure 28.

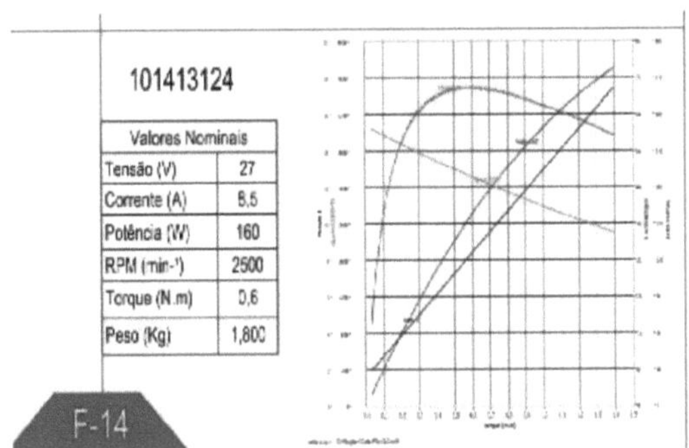

Figure 29- nominal motor data

Motor Torque (T1) = 0.6 N.m;

Pulley 1 diameter (d1) = 0.03 m;

Diameter of pulley 2 (d2) = 0.15 m;

Diameter of pulley 3 (d3) = 0.08 m;

Pulley Diameter 4 (d4) = 0.48 m;

Conversion from rpm to Hz

$$f_1 = \frac{2500}{60}$$

$$f_1 = 41{,}67\ Hz$$

Calculating the chair's final speed

$$V_1 = V_2$$

$$2.\pi.r_1.f_1 = 2.\pi.r_2.f_2$$

$$2.\pi.0{,}015.41{,}67 = 2.\pi.0{,}075.f_2$$

$$f_2 = 8{,}34 Hz$$

$$\omega_2 = 2.\pi.f_2$$

$$\omega_2 = 2.\pi.8,34$$

$$\boxed{\omega_2 = 52,34 \, rad/s}$$

$$\boxed{V_3 = \omega.r_3}$$

$$V_3 = 52,35.0,04$$

$$\boxed{V_3 = 2,093 \, m/s}$$

$$V_3 = V_4$$

$$V_4 = 2,09 \, m/s$$

Or

$$\boxed{V_4 = 7,5 Km/h}$$

12.2 - Calculating the chair's final torque

$$\boxed{\frac{D2}{D1} = \frac{T2}{T1}}$$

Where:

D2 = Diameter of the driven pulley;

D1 = Diameter of the driving pulley;

T2 = Torque moved;

T1 = Motor torque;

Data:

T1 = 0.6 N.m

Pulley 1 diameter (d1) = 0.03 m;

Diameter of pulley 2 (d2) = 0.15 m;

Diameter of pulley 3 (d3) = 0.08 m;

Pulley Diameter 4 (d4) = 0.48 m;

$$\frac{0,15}{0,03} = \frac{T2}{0,6}$$

$$\boxed{T2 = 3\ N.m}$$

$$\boxed{T2 = F2.R2}$$

$$3 = F2.\,0,075$$

$$\boxed{F2 = 40\ N}$$

$$T3 = F2.R3$$

$$T3 = 40.0,04$$

$$\boxed{T3 = 1,6\ N.m}$$

$$\boxed{\frac{D4}{D3} = \frac{T4}{T3}}$$

$$\frac{0,48}{0,04} = \frac{T4}{1,6}$$

$$\boxed{T4 = 19,2\ N.m}$$

$$\boxed{T4 = F4.R4}$$

$$19,2 = F4.\,0,24$$

$$\boxed{F2 = 80\ N}$$

Note that the torque that was 0.6 N.m has increased to 19.2 N.m, generating a force of 80 N per wheel, which generates a total force of 160 N, the force that will propel the chair.

Based on the calculations shown, the pulley and belt system of the chair was built. Figure 30 shows the first set of pulleys and Figure 31 shows the second set of pulleys.

In the first set, a Poly V micro V belt, model GPJ 0600, from goodyear, was used, and in the second set, an AA belt was used, more specifically the A55, and the larger pulley of this set is the wheel rim itself.

This ratio allowed us to increase the torque of the chair, enabling it to be used with loads of up to 100Kg, a value measured in practice, and generated a maximum speed of 7.5 km/h, also confirmed in the practical tests.

Figure 30- First reduction to increase motor torque

Figure 31- Second relationship

It is very important to highlight the partnerships of Laboratory 14 with the industry, because they allow students to have contact with the companies that they will work for in the future, thus allowing prototypes to get off the paper and become a potential product. Figure 32 shows the prototype of the motorized chair in full scale.

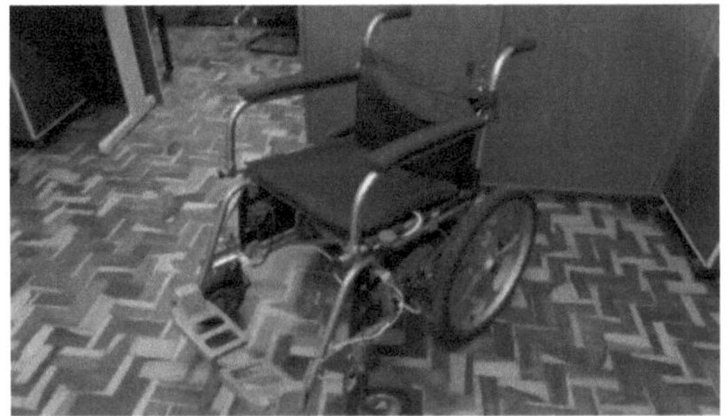

Figure 32- Full-scale motorized chair prototype

The mechanical structure was developed with metallic tubes, and the pedals and battery supports were made of cast aluminum to facilitate the construction in case the prototype can one day be built   in series.

The pedals, wheel supports, and battery support can be seen in figures 33,34, and 35.

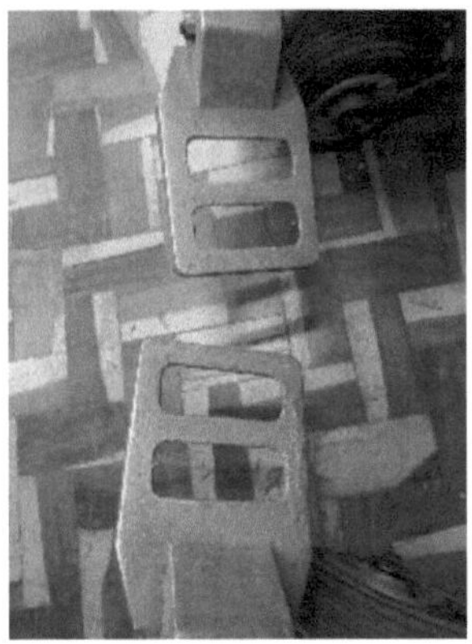

Figure 33- Pedals of the prototype, in aluminum.

Figure 34- Wheels Support.

Figure 35- Battery holder.

It is worth mentioning that the dimensions and protection systems of the prototype were based on the Brazilian technical standards (ABNT - ISO 7176-21:2009). Figures 36 and 37.

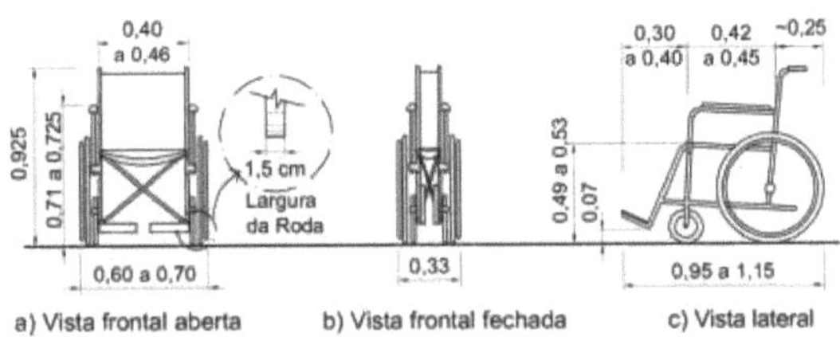

Figure 36- Limit Dimensions of the chair according to ABNT -2004 in meters

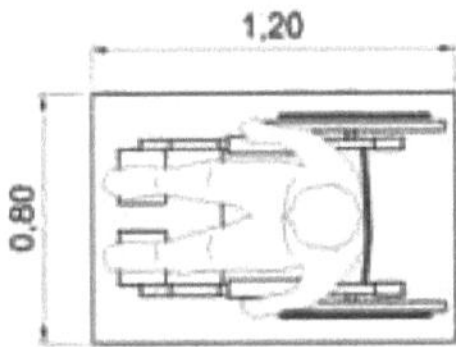

Figure 37- ABNT reference module - 2004 in meters

13 - Experimental Procedures

For safety reasons the first tests were performed inside lab 14, with the rear wheels up. The tests with the wheels touching the ground were only made after the logic errors in the *software*, small problems in the hardware, and the addition of a punch-type emergency switch (in case of emergency you only need to press it to interrupt the electrical connection between the batteries and the rest) were corrected. These tests were performed in the hallway of the Electronics Course due to limited lab space. Some volunteers helped with the tests. The source code that was written to the microcontroller is in Appendix A.

To make comparative tests, a joystick control system and an *Android* application were developed. Its source code is in appendix B.

14 - Conclusions and Recommendations

After several tests, the assembly showed a good result. Most commands spoken by the user were recognized correctly by the module. Since the user recorded his commands himself, the module does not recognize someone else's voice, which increases the system's security a little. However, in noisy environments, especially where there are many people talking at the same time, the module cannot differentiate the voice of the owner from the others, thus not recognizing the command. Cases of voice alteration, such as hoarseness or flu, also confuse the module, making it necessary to re-record the commands in these situations to allow its proper operation.

The voice module, which we use, also limits us to only five short commands (forward, reverse, right, left and stop), which limits the flexibility of the command mode. If it allowed us to record more commands, we could change the speed at which the chair moves without having to modify either the *software* or the *hardware*.

These problems could be corrected by using a more efficient speech recognition system, such as the one in Android that recognizes millions of words. Unfortunately, however, this only works online and has a high response time.

Now our next step will be to look for a voice control system with better performance, even if its cost is a little higher. This way it will become a viable system to go to the streets, without offering risks to the physical integrity of the user and others around him.

With the addition of Bluetooth control, the wheelchair user now has two modes of system control, which led us to think, in the next project, of incorporating a third control mode, by joystick, giving the motorized wheelchair user a good number of control modes, allowing the user to choose the one that best suits his or her disability.

APPENDIX - A (Voice Control)

```c
#include <16F877A.h>
#device ADC=10
#FUSES NOWDT //No Watch Dog Timer
#FUSES NOBROWNOUT //No brownout reset
#FUSES NOLVP //No low voltage prgming
#use delay(crystal=20000000)
#use rs232(baud=9600,xmit=PIN_C6,rcv=PIN_C7)
#include<lcd_galli.c>
//entries:
#define lock PIN_D3
#define brake PIN_D2
//exits:
#define Brake1 PIN_D1
#define Brake2 PIN_E2

char receives;
int1 flag=0;
int variance=1;
signed int16 pwm1=498,pwm2=498,final1=498,final2=498;
#int_TIMER1
void TIMER1_isr(void) {
signed int16 final1a,final2a,delta;
delta=(pwm2-pwm1)-(final2-final1);
final1a=final1+delta;
final2a=final2-delta;
if(final1a<50) final1a=50;
else if(final1a>947) final1a=947;
if(final2a<50) final2a=50;
else if(final2a>947) final2a=947;
```

```
if(final1a>(pwm1+variance)) pwm1+=variance;
else
if(final1a<(pwm1-variance)) pwm1-=variance;
else pwm1=final1a;
if(final2a>(pwm2+variance)) pwm2+=variance;
else
if(final2a<(pwm2-variance)) pwm2-=variance;
else pwm2=final2a;
```

```c
set_pwm1_duty((int16)pwm1);
set_pwm2_duty((int16)pwm2);
if(pwm1==498 &&
pwm2==498){
output_high(brake1);
output_high(brake2);
}
else{
output_low(brake1);
output_low(brake2);
}
set_timer1((int16)5553
6);
}
```

```c
#int_RDA
void RDA_isr(void) {
flag=1;
}
void main(){
output_high(brake1);
output_high(brake2);
lcd_ini();
printf(lcd_writes, "\fVoice Command version 2.7"); //indentify the
setup_ccp1(CCP_PWM) version;
setup_ccp2(CCP_PWM);
set_pwm1_duty((int16)498); //max =997
set_pwm2_duty((int16)498); //meio=498
setup_timer_1(T1_INTERNAL|T1_DIV_BY_1);//65.5 ms overflow
setup_timer_2(T2_DIV_BY_1,249,1); //50,0 us overflow, 50,0 us
interrupt
setup_comparator(NC_NC_NC);
enable_interrupts(INT_RDA); //received data via serial
enable_interrupts(INT_TIMER1); //end of timer 1 count
enable_interrupts(GLOBAL);
set_timer1((int16)0);
putc(0xAA); //initialize the voice module
putc(0x37); //activates compact mode
delay_ms(1000);
putc(0xAA); //initialize the voice module
```

```c
putc(0x22); //activate bank 2
while(TRUE){
if(!input(brake)){
final1=498;
final2=498;
variance=10;
}
else{
if (flag){
flag=0;
gets = getc();
switch(receive){
case '#': //pare
final1=498; //50% of pwm, 0% of speed
final1=498; //50% of pwm, 0% of speed
variance=10;
Printf(lcd_writes,"\f\b stop");
break;
case '!'://front
set_pwm1_duty(750); //front at 50% of speed
set_pwm2_duty(750); //front at 50% of speed
variance=1;
Printf(lcd_writes,"\f\b front");
break;
case ""://ready
final1=250; //is at 50% of the speed
final1=250; //is at 50% of the speed
variance=1;
printf(lcd_writes,"\f\b ready");
```

```c
break;
case '$'://right
final1=625; //front at 25% of speed
final2=375; //is at 25% of speed
variance=1;
Printf(lcd_writes,"\f\b right");
break;
case '%'://left
output_low(brake1);
output_low(brake2);
final1=375; //is at 25% of speed
```

```
final2=625; //front at 25% of speed
variance=1;
printf(lcd_writes,"\f\b left");
break;
}
}
}
}
}
```

APPENDIX B - Joystick control and Android application

```c
#include <16F877A.h>
#device ADC=10
#FUSES NOWDT //No Watch Dog Timer
#FUSES NOBROWNOUT //No brownout reset
#FUSES NOLVP //No low voltage prgming
#use delay(crystal=20000000)
#use
rs232(baud=9600,parity=N,xmit=PIN_C6,rcv=PIN_C7,bits=8,stream
=PORT1)
#include<lcd_galli.c>
//entries:
#define lock PIN_D3
#define brake PIN_D2
//exits:
#define Brake1 PIN_D1
#define Brake2 PIN_E2
int i=0,a[4],variance=1,j=100;
signed int16
b1=498,b2=498,bt1=99,bt2=99,js1=498,js2=498,adcx,adcy,pwm1=4
98,pwm2=498,
final1=498,final2=498;
#int_RDA
void RDA_isr(void) {
a[i]=getc();
if(a[i]==255)
i=1;
```

```
else{
if (i==3){
if(a[3]==244){
bt1=a[1];
bt2=a[2];
i=0;
}
}else
if(a[i]==233){
bt1=99;
bt2=99;
}else i++;
} 17
```

```c
if(i>3)
i=0;
}
#int_TIMER1
void TIMER1_isr(void) {
signed int16 final1a,final2a,delta;
delta=(pwm2-pwm1)-(final2-final1);
final1a=final1+delta;
final2a=final2-delta;
if(final1a<50) final1a=50;
else if(final1a>947) final1a=947;
if(final2a<50) final2a=50;
else if(final2a>947) final2a=947;
if(final1a>(pwm1+variance)) pwm1+=variance;
else
if(final1a<(pwm1-variance)) pwm1-=variance;
else pwm1=final1a;
if(final2a>(pwm2+variance)) pwm2+=variance;
else
if(final2a<(pwm2-variance)) pwm2-=variance;
else pwm2=final2a;
set_pwm1_duty((int16)pwm1);
set_pwm2_duty((int16)pwm2);
j++;
set_timer1((int16)55536);
}
void main(){
output_high(brake1);
output_high(brake2);
```

```c
lcd_ini();
printf(lcd_writes, "\fbluethoth + Joystick + version: 1.2"); //identifies
the software version
setup_ccp1(CCP_PWM);
setup_ccp2(CCP_PWM);
set_pwm1_duty((int16)498); //max =997
set_pwm2_duty((int16)498); //meio=498
setup_adc_ports(AN0_AN1_AN3);
setup_adc(ADC_CLOCK_INTERNAL);
setup_timer_1(T1_INTERNAL|T1_DIV_BY_1); //65.5 ms overflow
setup_timer_2(T2_DIV_BY_1,249,1); //50,0 us overflow, 50,0 us
interrupt
setup_comparator(NC_NC_NC); 18
```

```c
enable_interrupts(INT_RDA);
enable_interrupts(INT_TIMER1);
enable_interrupts(GLOBAL);
set_timer1((int16)0);
while(TRUE){
if(!input(brake)){
if(pwm1==498 && pwm2==498){
output_high(brake1);
output_high(brake2);
}
final1=498;
final2=498;
variance=10;
bt1=99;bt2=99;
js1=498;js2=498;
}
else{
output_low(brake1);
output_low(brake2);
if(!input(lock)){
set_adc_channel(0);
delay_us(10);
adcx=read_adc();
set_adc_channel(1);
delay_us(10);
adcy=read_adc();
if(adcx>498)
if(adcx<525) adcx=498;
else adcx-=27;
```

```c
if(adcy>498)
if(adcy<525) adcy=498;
else adcy-=27;
js1=(adcy+adcx)-498;
js2=(adcy-adcx)+498;
final1=js1;
final2=js2;
variance=1;
bt1=99;bt2=99;
}
else{
```

```c
b1=(bt1*5)+3;
b2=(bt2*5)+3;
final1=b1;
final2=b2;
variance=1;
}
}
if(j>100){ //writes on the display every half second
printf(lcd_writes, "\f%lu %lu %lu %lu",adcx,pwm1,js1,final1); //a:%lu
p1:%luna:%lu p2:%lu
printf(lcd_writes, "\n%lu %lu %lu %lu",adcy,pwm2,js2,final2);
j=0;
}
}
}
```

APPENDIX C - Control Board Component List

List of control board components

R1	10KΩ	**C3**	15pF
R2	50KΩ pot	**C5**	100nF
R3	820Ω	**C6**	1000µF / 25V
R4	820Ω	**C7**	100nF
R5	10KΩ	**C8**	100nF optional
R6	10KΩ	**C10**	1000µF / 16V
R7	2K2	**C11**	39nF optional
R8	2K2	**C12**	1000µF / 16V
R9	2K2	**Q1**	XTAL 20MHz
R10	2K2	**Q3**	XTAL 35KHz OPTIONAL
R11	560Ω	**CI 1**	PIC 16F877A OR EQUIVALENT
R12	10KΩ	**CI3**	PCF 8583 0PCIONAL
R18	10KΩ	**CI4**	24C01AP OPTIONAL
T1	2N2222	**D5**	1N4007
T2	2N2222	**D6**	1N4007
IC1	7806	**D7**	1N4148
TC2	7815	**D8**	1N4148
C2	15 pF	**D9**	1N4007

APPENDIX D -Component List Power Board

Component List Power Board

R2	820Ω - 1/8W	R23	820Ω- 1/8W	D7	1n4744 - 1w
R3	450Ω- 1/8W	R24	1kΩ- 1/8W	D9	1n4744 - 1w
R4	10kΩ- 1/8W	C1	100nF	D10	1n4744 - 1w
R5	10kΩ- 1/8W	C3	100nF	D11	1n4744 - 1w
R6	10kΩ- 1/8W	C4	1µF / 16V	D12	1n4744 - 1w
R7	10kΩ- 1/8W	C5	1µF / 16V	D13	1n4744 - 1w
R8	820Ω- 1/8W	C6	100nF	D14	1n4744 - 1w
R9	10kΩ- 1/8W	C9	100nF	D15	1n4744 - 1w
R10	10kΩ- 1/8W	C10	1000µF/25V	D18	1n4007
R11	820Ω- 1/8W	C11	1000µF/ 25V	CPTE2	4700µF / 50V
R12	10kΩ- 1/8W	C12	1000µF/25V	T1	2N2222
R13	22Ω- 1/8W	C13	1000µF/25V	T2	2N2222
R14	22Ω- 1/8W	C14	100nF	T3	2N2222
R45	820Ω- 1/8W	D1	1N4148	T4	2N2222
R16	10kΩ- 1/8W	D2	1N4148	T5	2N2222
R17	22Ω- 1/8W	D3	1N4148	Fet 1	IRF 3205
R18	10kΩ- 1/8W	D4	1N4148	Fet 2	IRF 3205
R19	22Ω- 1/8W	D5	1N4148	Fet 3	IRF 3205
R20	10kΩ- 1/8W	D6	1N4148	Fet 4	IRF 3205
R21	10kΩ- 1/8W	D16	1N4148	CI1	IR 2111 or IR 2184
R22	1KΩ- 1/8W	D17	1N4148	CI2	IR 2111 or IR 2184
LED	5mm	Terminal blocks	2 pins	2 sockets	8 pins
IC3	7815				

Bibliography

AHMED, Ahmed. **Power Electronics.** São Paulo:Prentice Hall, 2000.

MALVINO, Albert Paul. **Eletrônica.** São Paulo: Makron Books, 1997. V. 1;

PEREIRA, Fábio. **PIC Microcontrollers - Programming in C.** 7th Ed. Editora Érica, 2009.

SOUZA, David José de. **Desbravando o PIC - Extended and Updated for PIC 16F628A.** 12th Ed. São Paulo: Érica, 2008

NOBORU, A. PIC18 Microcontrollers: Learn and Program in C Language. 1st ed. São Paulo: Editora Érica, 2009.

SILVA, R. Programando Microcontroladores PIC em C: Com base no modelo PIC16F628A. 2ª ed. São Paulo: Ensino Profissional, 2006.

BRAGA, C. Basic Electronics for Mechatronics: Electronics, Mechanics and Mechatronics. 1st ed. São Paulo: Saber Editora, 2005.

SOUZA, V. Programming in C for the AVR: Fundamentals. 1ª ed. São Paulo: Professional Teaching 2011.

MONTESSORI, M. Para Educar o Potencial Humano. Papirus, 2014.

https://www.techtudo.com.br/artigos/noticia/2012/01/bluetooth-o-que-e-e-como-funciona.html

http://www.abnt.org.br/noticias/4375-cadeira-de-rodas#:~:text=This%20Part%20of%20ABNT%20NBR,to%20subscribe%20and%20descend%20obst%C3%A1cles.

https://www.abntcatalogo.com.br/

http://abnt.org.br/paginampe/biblioteca/files/upload/anexos/pdf/d061552995d91816081c20f05ae713f4.pdf

ABNT 9050. Accessibility to buildings, furniture, urban spaces and equipment. ABNT - Brazilian Association of Technical Standards. Rio de Janeiro, p. 97. 2004.

Printed by Books on Demand GmbH, Norderstedt / Germany